Fighting With Your Eyes Closed:

A First Year's Guide to Medical School

Jelani D. Ingram

ISBN 978-0-6151-4331-6

For more information, contact Jelani D Ingram at
jdingram@gmail.com

Contents

To   Alton and Nancy Ingram
        Milton and Gladys Pettis

Topic 1:
In The Beginning

The honest truth is that most of us are smart…well, okay all of us are smart. However now entering a region unknown to many, most who read this will be an average student for the next four years of their life. The previous statement is not indicating that you will become less intelligent, but harshly describing the reality of medical school. Now let's focus on the hard part of school, yes there is a hard part other than the work.

Did you ever wonder what is the right thing to say during the interview question: "What do you do for fun?" Did you ever just wonder why they ask that question? "Social Interaction" is the answer. Medical School is not hard just because of the work; it's hard because you have to "live". Let me rephrase that. It's hard because it's almost impossible to just focus on your school work and not spend time living, but still be able to be human after you finish. I will say it clearly for the unlucky few who are not familiar with the language of sarcasm. It's hard to have a normal life and study the amount of work given within each semester. I don't believe doctors get their degree for finishing the required work, I think it's because medical school assaults them with a giant hammer and they miraculously lived.

Bear with me for awhile and we will uncover the path to an easier first year of medical school. Depending on how your attention deficit is treating you… you might even finish the book in one sitting! Let's GO!

Topic 2:
The Work

This will be shorter than expected, because there are different styles for different people. Though there are a few things I could say to help assist your voyage. First things first: The Syllabus. In college, most of us received the syllabus and it is lost within a day or so. Then we try to find it when the professor states, "You didn't know… it's in your syllabus?" The days of having a syllabus of 3-5 pages and gave you the professors' horoscope… are over. Difference A: your syllabi cost money and not just five dollars if you get my drift. Difference B: your syllabi can range anywhere from 100 to 300 pages per class. So if your school likes "classical": that is four classes' multiplied by an average of 200 pages covered in four weeks. If your school likes "modern rock": that is one class multiplied by an average of 250 pages covered in about a week and a half. Notice: Don't kill the messenger, I'm not responsible for your jaw dropping. Your syllabus is a "per class bible". You will need this to study for the class and protest any mistakes made by the professors on the test.

Second: School Supplies. I was standing in the store before my first semester, next to a young girl picking her nose and her older brother trying to eat a glue stick, when I realized that there was not many things I need for school except my brain. Although, you will need two vital supplies: highlighters and pencils. The highlighters are for you to cover your syllabus like a rainbow every time one is handed to you. The pencils are for your very fun examinations. This is the big leagues! The NFL has professional football players, Wallstreet has professional business people, and medical school has professional test takers. Everyone, like in the NFL, is not the star but we are all there as professionals.

Third: Studying and Test Taking. I hope you didn't think the MCAT was hard... ha ha! Studying A: first try to identify if the class is raw memorization (Microbiology), Conceptual (Physiology), or a mix of both (Anatomy). Studying B: then study accordingly... that's it! I know you're thinking that can't be it. Well in a nutshell it is, but the "art" of studying, I will let you uncover. There are different types of students out there, for example: Students who finish the test thinking they did horrible and end up earning a good grade and those that finish the test thinking they did horrible and they're right… they did horrible. The last point worth mentioning is the measurement of your study practices. If you get to the test and it seems like the professors had a couple of drinks, tied a blindfold around their eyes', and selected random sentences from your syllabus… well it may be time to change your study habits.

Melissa Darjon, University of Texas Southwestern Medical Center at Dallas

"The work…a lot of volume in a very short period of time!"

April Farley, Baylor College of Medicine

"The work is something that you couldn't explain…it just has to be experienced."

Topic 3:
Money Cents

If you didn't know... you don't make money while in school, but you do spend a great deal of it. So where does it come from and where does it go? Welcome to the wonderful world of financial aid: loans and grants. In medical school, unfortunately, it's more loans obtainable than grants. To make loan money work, you need the following: Great Decision Making!

If you were down to your last fifteen dollars would you buy a medical review book or a twenty-four pack of ice cold, mouth watering, delicious beer? For those of you that answered beer, you just took the wrong way at the fork in the road. Good decision making with loan money involves planning most of all, because all of the loan money comes at the beginning of each semester. If you buy everything you have ever wanted for your new residence, during your first semester, at final exam time it will be Spam, Vienna sausages, and Ramen noodles for you! Considering students need all the nutrients they can get during exams, no money is a BAD thing. So for those of you who want a small amount of guidance, I've included a financial worksheet.

This worksheet is by monthly payments (Rent or mortgage could be from $500 to $1200, therefore I will use $720 to make it simple.)

Monthly Payments

House

Rent	$500-1150
Electric	$70-130
Gas (if applicable)	$10-20
Phone Bill (house)	$25-50
Phone Bill (cell)	$40-70
Water	$7-20
Cable	$30-70
Internet	$17-50

Car

Car Wash	$5-40
Gas (not optional)	$75-120

Personal

Food	$150-250

-Example-
$50 for one week of groceries w/THREE meals a day= $200/ month (ALL THREE meals)

Eating out TWO meals a day, for 7 days a week at 5 dollars w/ no tax= $280/ month (Just TWO meals)

Haircut or Style	$24-55

Random (unexpected)

Magazines	$0.50-2
DVD/ Movies	$8-40
Bank Services Charge	$0-5
Alcohol/ Bars	$10-100
Impulse Buying!	$10-50

Average

Minimum Total $981.50
Maximum Total $2,222.00

*Remember these are estimate averaged totals, may be different for NY or L.A.

Hints: 1) Know your credit score- this will help out in the long run when you want to make a large purchase. 2) Some students find it useful purchasing a condo- (this is not for everyone) if you rent for four years you would spend about $37,000 depending on where you live in America. Buying is useful because even if the value of the condo decreases you still wouldn't lose as much if you paid for an apartment for four years. For example, if you spent $40,000 during those four years and you sold for $20,000 less than what it was bought for, then you would only lose about $20,000 instead of $40,000. Again, this method of saving money is not for everyone! Check with the real estate advisors in your region to determine if this is right for you. 3) Some students use their "rewards" credit card to pay for their tuition- (the process of payment differs from school to school) this strategy is useful because the money that is usually sent straight to the school for tuition, will now be sent to your personal account in order to pay off the credit card. This process doesn't affect your loan money; it just gives you the opportunity to take a vacation, just for paying your tuition. I know… it's a sweet deal! 4) Buy materials like paper towels and toilet tissue in bulk- this may seem obvious, but so many of us forget to save money this way. 5) Know special situations with your financial aid office- many financial aid offices extend your loans for expensive car maintenance and also reimburse your payment of a brand new, shiny computer. Yeah…I know, just thank me later when you are not struggling at the end of your semester. For those of you who live like rock stars during the semester, there are usually emergency loans that don't hurt your pockets too hard.

Just remember to make a flow chart after the first month and monitor your spending each month. Try your best to stick to your allowance each month. I know one student who keeps two accounts: one for monthly spending and the other to hold the rest of the loan money to gain 5% interest. The student simply transfers his personally made monthly stipend into the first account at the beginning of every month. With every tip mentioned, just be smart with your money (that you haven't made yet).

Sheena Heslop, University of Medicine of New Jersey

"If you live like a doctor when you're a student, you'll live like a student when you're a doctor."

Ahmad Garrett-Price, University of Texas Southwestern Medical Center at Dallas

"Budgeting is something that I am still figuring out…now I categorize my finances so I know where I am spending my money!"

Vincent Scott, University of Texas Medical School at Houston

"Since I didn't enter medical school straight from college, it was a big adjustment…I was accustomed to having a constant income. If you come straight from college it's easier because you are accustomed to having no money!"

Topic 4:
Living Healthy

A two inch deep skillet filled with the biggest chocolate chip cookie, covered with three huge scoops of vanilla ice cream, which is also covered with steaming hot melted chocolate syrup… sounds so delicious. Now slap yourself on the hand because that's not healthy, although it is sometimes necessary in order to forget all your mental stress. Eating right is unfortunately not the only aspect to healthy living. For all you insomniacs out there…this is going to hurt…sleeping enough hours may be, arguably, the most important factor of a healthy life. When a person obtains enough hours of sleep they are able to focus on his/her class material the next day. The person is also more willing to exercise during the day and more inclined to cook because there is no need to "save time". Now I know in this new world some of our future female doctors, lawyers, and homemakers don't know how to cook, but getting enough hours of sleep (at least) allows them to experiment in the forbidden garden of Kitchenware. I will be the first one to tell you that this sleep issue is going to be different across schools. Why? Well because, of the transition of some schools from the Traditional teaching method (of having subject based learning i.e. Anatomy, Microbiology) to System based learning (i.e. learning everything about the cardiovascular system). The Traditional students will find it a little more challenging to get the required amount of sleep during test time, if their school teaches more than two subjects at once. Therefore, it is extremely important to know when your tests are and plan accordingly.

If you only run 20 miles a day you'll be fine... and you will also be ready for the Olympics! But realistically, exercise does wonders for your attitude, self-esteem, and not to mention your weight. During exercise, a decrease in pH, and an increase in 2,3 BPG shifts the Bohr curve to the right...just kidding (you'll get enough of that in school). I know it is sometimes hard to make yourself go to the gym, it happens to everyone from time to time. What keeps me at the gym is my six week program and diversity when working out. I conduct a six week program with defined goals. Anything longer than six weeks I began to lose interest. I set the goals that I want to achieve at the beginning of each week, that way I actually see myself improving. Setting these goals doesn't take as much work as it may seem, take the first week to "test the waters" and gain knowledge of what you will be able to handle. During the bulk of the six week period keep to your regimen and workout about 3-4 times a week depending on your schedule. I recommend skipping one day between your workouts the first couple of weeks to get your body accustom to your new regimen.

Make sure to add diversity to your exercise schedule! Diversity of activities like swimming some days or running outside instead of inside will keep you motivated. To assist your motivation, add at least one diverse day to your schedule during the week. After I finish my six week schedule, I usually take about one week off. This week off includes eating healthy and sleeping right, so I will not jeopardize my success. Here are a few examples of diverse workouts:

-Swimming
-Running on different trails and paths outside
-Tennis/Racquetball
- Basketball
-Running on stadium bleachers
-Yoga/Pilates
-Indoor/ Outdoor Aerobics

Tips:
Men- To gain brut strength in less time, space your repetitions 2 minutes apart.
Women- to get that "perfect stomach", you have to lose total body weight (Just doing sit-ups will not cover it).

Now let's rewind to eating! Cooking to me and some of my friends is fun and exciting; to some of my other friends it's scary and extensive. Cooking has two "angel wings": 1) it saves money and 2) it makes a person healthier. Even if you think you can't cook, the internet is so advanced now that you can get free recipes on the internet. Some of the recipes are hard, but many of them are easy and only require the mind set of an 8 year old. Remember if an ingredient is listed that you are not familiar with, use a search engine to look for an image. Fast food is a nemesis? No, not really. I know the health freaks are now, online trying to find my address to toilet paper my house for making that statement. However, sometimes… time is the nemesis. So in these situations take your life position in consideration- meaning if you are on your rigorous workout schedule, a burger and fries won't hurt as much as if you haven't exercised in three weeks. Many fast food places have nutritional information about their meals…ask for them. Everything said and done, just know what you are eating. Too much of anything is dangerous to your health, so pay close attention in Biochemistry and Physiology!

Tony Rozier, Boston University School of Medicine

"It gets crazy sometimes but you always schedule in eating so you can schedule in working out."

Ahmad Garrett-Price, University of Texas Southwestern Medical Center at Dallas

"I am a former athlete, so I make it a point to eat right… that's a personal decision."

Topic 5:
Free Time

There is not that much free time in medical school! Sorry, I know you were a little bit disappointed in the last sentence's lack of enthusiasm. But being in medical school is all about balancing the time you have to study. I know what you're thinking, "I know it's all about time management." But it…really is. For example, right now it is 11:00pm, I am at school, my Microbiology book is laying to my right, and I decided to turn to my computer and write about "free" time. Well, I know what you are thinking, "Man…that guy is really a nerd." My response, "Yeah I know." Guess what if you are going to medical school you are too. We are the world's collection of the coolest nerds, but nerds nonetheless!

Free time is what you make it in medical school. There are definitely times when my friends and I have had fun with our free time. Medical school would be easier if there was a way to turn off the emotional part of it. Utilizing free time is extremely important for the emotional part of school. There is actually more free time than you think (once you become adjusted to the workload). So loneliness becomes a factor, once you notice your free time. Try your absolute best not to become lonely, because loneliness is sometimes accompanied by bad habits. If you didn't know, there is said to be a high prevalence of physicians with substance abuse problems. So be very conscious about the regulation of your mental and physical health. You as a physician can only help someone so much before your own problems start to interfere.

A good old-fashion trick is: take up some kind of hobby when you get there. You are spending four of your years learning, you might as well add on something fairly unrelated. My girlfriend and I were joking around the other day and she thought I was so nerdy for writing a book in the middle of school. However, this is just something that takes my mind away from acid-base renal physiology (and you thought acid-base was over). Disclaimer: I am not instructing you to obtain a dangerous hobby. If you want to sky dive in you free time that is totally up to you. Don't come back crying, saying that I told you to collect swords and your hand got chopped off. Realistically, choose something that doesn't require the company of another person and don't be embarrassed about what it is. Your hobby…is yours!

Sam Wyche, Temple University School of Medicine

"When I'm not on "rounds", I'm reading. When I'm not reading, I'm with my girlfriend…she takes up all my free time!"

Robert R. Drummond, Johns Hopkins
School of Medicine

"I picked up a hobby…working on cars
calms me down on the weekend and takes
my mind off of school."

Topic 6:
Family Matters

Congratulations, you are now the NEW family physician so get accustom to the questions! You all now have joined the elite, who are family physicians four years before they graduate. I remember the day I got in the medical school, my friend told me he had a sore throat and asked me to look at it. I specifically remember the correct clinical presentation of his throat: hot and slimy. You're not going to really know anything until your 3rd year, so make sure you pass the memo on to your family. All of your friends, who are not in graduate schools, will have an idea of what you are doing, however they will not know the depth of your new acquired lifestyle. There will also be people who ask you how medical school is going. You say, "Fine!" then they will ask you, "What are you MAJORING in?" Don't frown up like a toddler took his diaper off and threw it in your face. Just smile and say, "I'm studying to become a doctor, but I still have time to decide what type." Many, many people (more than you can imagine) have no idea what medical school really is. Some don't even know that medical students have their bachelor's degree. So be polite and just understand that ignorance (of anything) doesn't mean stupid, but a lack of knowledge!

FYI: In this book when I mention the word friends, I am referring to your close friends. I am referring to the friends that you would trust with your well-being, friends that you would trust with your mother's well being, or friends that you would trust with your child's well being. From a totally different point of view your family and close friends will be your key to unlocking the mental strength needed to survive! There was this one girl I knew…let's called her "Girly". Girly came into medical school extremely determined like the rest of us! However, Girly had a gift that consisted of being able to block out every distraction including her family. She could spend the night at the school for 72 hours with an inflatable mattress, coffee maker, and sandwich meat. Girly could also not talk to her family and keep her head down when passing by a fellow classmate. Her carefully designed plan to excel in school worked perfectly when test time came around. Girly was one of the top students in her class; she always seemed to set the curve. After the third test block came around she decided that she would celebrate, because of her success in her courses! On Wednesday evening she decided to cook a tasty French cuisine. While she chopped and prepared her dinner, she went through her list of phone numbers to call a friend to plan how to celebrate the occasion. Although, going through her phone she noticed that she had not received very many phone numbers from her classmates. Of the two numbers she tried one was disconnected and the other didn't answer. Girly didn't leave a message because she was ashamed to ask the person to call her back.

Later that evening, after dinner, Girly decided to pick up a bottle of wine and a movie (probably Hollow Man 2 or Bring it on Again!) As she drank the wine she became more and more light headed, but she also felt free to continue drinking. "The

night was a beautiful bright blue ocean",
Girly thought as she walked outside to
smell the fresh air. That was Girly's very
last thought… before she woke up at her
schools teaching hospital with a very
nervous third year medical student staring
at her charts. "You fainted outside your
house and two of your drunken classmates
found you", the third year quietly uttered.

Don't drink and look at the stars is
the moral of this story. Not really! The
moral is not to outcast the people who love
you; utilize their compassion for your
success. Remember, your family and close
friends will be your key to unlocking the
mental strength needed to survive!

Monica Cantu, University of Texas
Medical School at Houston

"Family is important...sometimes I have to
wait until I have free time because of my
busy schedule."

Vince Scott, University of Texas Medical School at Houston.

"I talk with my parents two to three times a week…If a big family problem arises, that is more important to me than any one test."

Topic 7:
Wilder Partiers

Do you remember your craziest college party? Warning: don't think out loud if your mom, girlfriend, boyfriend, or grandparents are close. SHH! We will cognitively whisper during this section. College parties are…well they are…let's just say they are not for the light-hearted or the children. The craziest, most absurd, shocking things could be viewed at one of these events. So what makes first semester medical students party harder? I conducted a scientifically accurate survey where I randomly selected 5 of my own friends who think just like me. Drum roll please….

Second runner-up: "It's the last time to party before we really get serious". Most senior college students repeat the same phrase right before they graduate. However, medical students realize that they were granted one more opportunity to "party like its 1999" (sorry I just couldn't resist).

First runner-up: "We poor medical students don't get out that much". When I say poor I mean both the literal and figurative definition. Sorry future doctors, during those four years you sell your soul to your selected school. It's kind of like the NFL draft… except you give the school your signing bonus of 130,000 dollars that you have not made yet. Yeah… sounds like fun already!

Winner by knock out: "I deserve to party, I just learned a college semester in three and a half weeks". This is by far the most used excuse for partying hard. After the last question is answered on your "after the exam" survey of your first test or test block, you and your pre-doctor friends hurry home, not to sleep but to start celebrating early. Sleeping is for the next day when you morph into Rip Van Winkle.

On a more serious note, celebrating can be extremely fun and stress relieving, but binge drinking can be extremely dangerous. Many students end up in the hospital because they don't understand that the body has a limit. The liver can only filter out a limited amount of toxins from your body at one time. Let's break your test week down. Lots of coffee, no sleep, stressed out, unhealthy eating, and your body hasn't handled alcohol all week. It doesn't sound like a very good defense for alcohol poisoning. In addition, be very conscious of your ability to drive home after a "night at the town". So many people die from drunk driving every year, its ridiculous. So please, please be careful when celebrating a victory at school. The victory is no good if you are dead!

Sumiko Armstead, Baylor College of Medicine

"It seems when people get to med school, they are already experienced drinkers... in college, people are still experimenting."

Jason Sutton, University of Texas
Medical School at Houston

"Sometimes parties and drinking is needed
in order to get away and talk about things
other than medicine!"

Topic 8:
Girlfriend, Boyfriend, or Not!

Have you ever set and talked to your boyfriend for four straight hours? Have you ever argued with your girlfriend until 6 o' clock in the morning? Better yet have you ever tried to study while you're angry at your loved one? If you answered yes to any of these questions, you've been a victim of what I call "Stress Diversion" or SD. The funny thing is that SD only happens when work needs to be done. During a heated discussion you inquire what in the hell your love one is talking about, at that point your school stress is diverted! The self proclaimed theory is that people use their love ones to divert their stress. Test week is the highest frequency of SD because students are quiet so long and thinking so hard. So the question becomes: How is SD controlled?

The first step is to decide if you want a boyfriend or girlfriend. The first semester will be difficult. At some point you will, more than likely, become lonely even if you attend school in your home town. You have to be mindful that nobody except doctors and other medical students know what you are going through. Not knowing exactly about the difficulty level will be important when you have to deal with someone outside of medical school. Even other students, not in medicine, don't fully understand our matriculation as we don't fully understand theirs. As a result, there will be misunderstandings about simple and trivial issues that may not resolve quickly. Try to think back to college and remind yourself if you are more efficient with or without a girlfriend (or boyfriend). I've noticed some people do terrible when in a relationship, because of the stress. Others do better in a relationship, because it forces them to organize their time. At my school it seems that married couples do a tad better than the average student. They do better coping with the over all mental stress of medical school.   If you find that you want a relationship, be particularly truthful with the person to eliminate various arguments. **Notice**: It also gives you the advantage during the argument because you already told them! Most of the time both you and the person will be stubborn during the argument until somebody becomes the "bigger person" and stops the ridiculous fight about having Dr. Pepper or Coca-Cola (the answer is RC cola).

> **Fact:** It is hard trying to maintain a relationship entering medical school.
> **Opinion:** The medical school gods play battleship with long relationships during your first year of school.

So are you trying to keep your boyfriend or girlfriend you've been dating for one year? Good luck! It seems that medical school doesn't care about the length of time the relationship has lasted. All school cares about is that you suffer during the first year, and the other person receives the same fate. To help as much as I can, I've included a list of 6 useful tips!

1. Don't get into an argument about one subject and bring up other subjects you dislike.
2. EX-factor: there is an ex some where in both persons lives that will be there until marriage. (this doesn't mean that there is a strong attraction)

3. Don't change yourself for the other person. Merge both parties' ideas and thoughts.
4. The best friend knows about your arguments. Don't assume they have no idea!
5. Learn from your past relationship, but remember everyone is different.
6. Guys, you will never win an argument!

David Jones, Boston University School
of Medicine

"People say there is no time for
relationships, but that's not true...you
have time for what you want!"

Brannon George, University of Texas Medical School at Houston

"Since she is in dental school and I'm in medical school…we will sometimes study together. Even though it may sound boring, at least we're together and around each other."

Lisa Williford, University of Texas Medical School at Houston

"I haven't found any negatives to being married…I wouldn't be able to get through med school without him, so sometimes I have to take time and not study to spend time with him."

Topic 9:
End of the Line

There is one last item on the itinerary, the statement "You'll be okay!" Get ready because you will hear this statement a total of 32 times which equal about how many test you will have your first year.

Users of statement: Upperclassmen/ Professors

Victims of statement: First years

Some use this statement every time they give advice and others when trying to calm down another person. I typically don't like to use the statement because it is like a curse word: it is only used when a person can't think of any other words to say. It does not really mean you will be okay, in reference to the test. It means you will be okay once you become a physician. So, don't worry…You'll be okay!

Congrats! You are now a pre-doctor. Your abilities don't even include changing bed pans. On the other hand, you are more prepared to handle you first year in medical school where you will gain a great deal of knowledge. There will be stress, family issues, and emergencies during your first year. Try to handle those issues with the best of your abilities, just don't panic. Keep in mind thousands of people have made it through medical school to become doctors and so will you. The admissions committee does a good job in finding the right candidates... THEY FOUND YOU!

Special Thanks!
To my pre-readers: Mom, Dad, my sister,
Aunt Becky, Nick, Fenwa, and Jon
To my friends who donated quotes!
To my first year study group… the loudest
group ever (Jason, Lindsey, Brannon,
Vince, Fenwa, and of course Marquita)
To my professors that help me: Dr.
William Seifert, Dr. Paula O'Neal, Dr. Len
Cleary, Dr. Rebecca Cox, and Dr. Han
Zhang
To: Dr. Charles Willis, Dr. Robin
Williams, Dr. Louis Jean-Moore, Mr. and
Mrs. Franklin, Mrs. Cammon, Mrs. Pat
Caver, Dr. Margaret McNeese, Senator
Royce West, Ms. Oprah Winfrey, DeSoto
Independent School District, Morehouse
College, and The University of Texas
Medical School at Houston!

Socialmedworld.com

This is a list of all the M.D. Programs in the USA

Alabama

- University of Alabama School of Medicine
- University of South Alabama College of Medicine

Arizona

- University of Arizona College of Medicine

Arkansas

- University of Arkansas College of Medicine

California

- David Geffen School of Medicine, UCLA
- Keck School of Medicine, University of Southern California
- Loma Linda University, School of Medicine
- Stanford University School of Medicine
- Western University of Health Sciences
- University of California, Davis, School of Medicine

- University of California, Irvine School of Medicine
- University of California, San Diego, School of Medicine
- University of California, San Francisco

Colorado

- University of Colorado Health Sciences Center School of Medicine

Connecticut

- University of Connecticut, School of Medicine
- Yale University School of Medicine

District of Columbia

- George Washington University School of Medicine and Health Sciences
- Georgetown University School of Medicine
- Howard University College of Medicine

Florida

- Florida State University College of Medicine
- Florida International University College of Medicine opening Fall 2009

- University of Florida College of Medicine
- University of Miami Miller School of Medicine
- University of South Florida College of Medicine
- The University of Central Florida, College of Medicine (established by the Board of Governors March 23, 2006)

Georgia

- Emory University School of Medicine
- Medical College of Georgia, School of Medicine
- Mercer University, School of Medicine
- Morehouse School of Medicine

Hawaii

- John A. Burns School of Medicine (University of Hawaii)

Illinois

- Chicago Medical School (Rosalind Franklin University of Medicine and Science)
- Feinberg School of Medicine (Northwestern University)

- Pritzker School of Medicine (University of Chicago)
- Rush Medical College (Rush University)
- Southern Illinois University School of Medicine
- Stritch School of Medicine (Loyola University of Chicago)
- University of Illinois College of Medicine

Indiana

- Indiana University School of Medicine

Iowa

- University of Iowa Roy J. and Lucille A. Carver College of Medicine

Kansas

- University of Kansas School of Medicine

Kentucky

- University of Kentucky College of Medicine
- University of Louisville School of Medicine

- Pikeville College School of Osteopathic Medicine

Louisiana

- Louisiana State University School of Medicine in New Orleans
- Louisiana State University School of Medicine in Shreveport
- Tulane University School of Medicine

Maryland

- Johns Hopkins University School of Medicine
- F. Edward Hebert School of Medicine at the Uniformed Services University of the Health Sciences
- University of Maryland School of Medicine

Massachusetts

- Boston University School of Medicine
- Harvard Medical School
- Tufts University School of Medicine
- University of Massachusetts Medical School

Michigan

- Michigan State University College of Human Medicine
- University of Michigan Medical School
- Wayne State University School of Medicine

Minnesota

- University of Minnesota Medical School
- Mayo Clinic College of Medicine

Mississippi

- University of Mississippi School of Medicine

Missouri

- Saint Louis University School of Medicine
- University of Missouri-Columbia School of Medicine
- University of Missouri-Kansas City School of Medicine
- Washington University in St. Louis School of Medicine

Nebraska

- Creighton University School of Medicine
- University of Nebraska College of Medicine

Nevada

- University of Nevada School of Medicine

New Hampshire

- Dartmouth Medical School

New Jersey

- UMDNJ-New Jersey Medical School
- UMDNJ-Robert Wood Johnson Medical School

New Mexico

- University of New Mexico School of Medicine

New York

- Albany Medical College
- Albert Einstein College of Medicine
- Columbia University College of Physicians and Surgeons
- Weill Cornell Medical College

- City University of New York Medical School
- Mount Sinai School of Medicine
- New York Medical College
- New York University School of Medicine
- State University of New York Downstate Medical Center College of Medicine
- State University of New York Upstate Medical University
- Stony Brook University Health Sciences Center School of Medicine
- University of Buffalo State University of New York School of Medicine & Biomedical Sciences
- University of Rochester School of Medicine and Dentistry

North Carolina

- Duke University School of Medicine
- The Brody School of Medicine at East Carolina University
- University of North Carolina School of Medicine
- Wake Forest University School of Medicine

North Dakota

- University of North Dakota School of Medicine and Health Sciences

Ohio

- Case Western Reserve University School of Medicine
- Cleveland Clinic Lerner College of Medicine
- Medical University of Ohio
- Northeastern Ohio Universities College of Medicine
- Ohio State University College of Medicine
- University of Cincinnati College of Medicine
- Wright State University Boonshoft School of Medicine

Oklahoma

- University of Oklahoma College of Medicine

Oregon

- Oregon Health and Science University School of Medicine

Pennsylvania

- Drexel University College of Medicine
- Jefferson Medical College of Thomas Jefferson University
- Pennsylvania State University College of Medicine
- Temple University School of Medicine
- University of Pennsylvania School of Medicine
- University of Pittsburgh School of Medicine

Puerto Rico

- Ponce School of Medicine
- San Juan Bautista School of Medicine
- Universidad Central del Caribe, School of Medicine
- University of Puerto Rico, Medical Sciences Campus

Rhode Island

- Warren Alpert Medical School of Brown University

South Carolina

- Medical University of South Carolina College of Medicine
- University of South Carolina School of Medicine

South Dakota

The Sanford School of Medicine of The University of South Dakota

Tennessee

- East Tennessee State University James H. Quillen College of Medicine
- Meharry Medical College School of Medicine
- University of Tennessee Health Science Center College of Medicine
- Vanderbilt University School of Medicine

Texas

- **University of Texas Medical School at Houston!**
- Baylor College of Medicine
- Texas Tech University Health Sciences Center
- Texas A&M System Health Science Center - College of Medicine
- University of Texas Medical School at San Antonio
- University of Texas Southwestern Medical School in Dallas
- University of Texas Medical Branch at Galveston

Utah

- University of Utah

Vermont

- University of Vermont College of Medicine

Virginia

- Eastern Virginia Medical School of the Medical College of Hampton Roads
- University of Virginia School of Medicine
- Virginia Commonwealth University

Washington

- University of Washington School of Medicine

West Virginia

- Joan C. Edwards School of Medicine at Marshall University
- West Virginia University School of Medicine

Wisconsin

- Medical College of Wisconsin

- University of Wisconsin School of Medicine and Public Health

www.ingramcontent.com/pod-product-compliance
Lightning Source LLC
Chambersburg PA
CBHW032015190326
41520CB00007B/486